UNITED STATES CIVIL SPACE POLICY

SUMMARY OF A WORKSHOP

Molly K. Macauley, Rapporteur
Joseph K. Alexander, Rapporteur

Space Studies Board
and
Aeronautics and Space Engineering Board
Division on Engineering and Physical Sciences

NATIONAL RESEARCH COUNCIL
OF THE NATIONAL ACADEMIES

THE NATIONAL ACADEMIES PRESS
Washington, D.C.
www.nap.edu

THE NATIONAL ACADEMIES PRESS   500 Fifth Street, N.W.   Washington, DC 20001

NOTICE: The project that is the subject of this report was approved by the Governing Board of the National Research Council, whose members are drawn from the councils of the National Academy of Sciences, the National Academy of Engineering, and the Institute of Medicine.

This study was supported by Contract NASW-01001 between the National Academy of Sciences and the National Aeronautics and Space Administration, and private funding from the National Research Council. Any opinions, findings, conclusions, or recommendations expressed in this publication are those of the authors and do not necessarily reflect the views of the agencies that provided support for the project.

International Standard Book Number-13: 978-0-309-12014-2
International Standard Book Number-10: 0-309-12014-4

Copies of this report are available free of charge from:

Space Studies Board
National Research Council
500 Fifth Street, N.W.
Washington, DC 20001

Additional copies of this report are available from the National Academies Press, 500 Fifth Street, N.W., Lockbox 285, Washington, DC 20055; (800) 624-6242 or (202) 334-3313 (in the Washington metropolitan area); Internet, http://www.nap.edu.

Copyright 2008 by the National Academy of Sciences. All rights reserved.

Printed in the United States of America

THE NATIONAL ACADEMIES
Advisers to the Nation on Science, Engineering, and Medicine

The **National Academy of Sciences** is a private, nonprofit, self-perpetuating society of distinguished scholars engaged in scientific and engineering research, dedicated to the furtherance of science and technology and to their use for the general welfare. Upon the authority of the charter granted to it by the Congress in 1863, the Academy has a mandate that requires it to advise the federal government on scientific and technical matters. Dr. Ralph J. Cicerone is president of the National Academy of Sciences.

The **National Academy of Engineering** was established in 1964, under the charter of the National Academy of Sciences, as a parallel organization of outstanding engineers. It is autonomous in its administration and in the selection of its members, sharing with the National Academy of Sciences the responsibility for advising the federal government. The National Academy of Engineering also sponsors engineering programs aimed at meeting national needs, encourages education and research, and recognizes the superior achievements of engineers. Dr. Charles M. Vest is president of the National Academy of Engineering.

The **Institute of Medicine** was established in 1970 by the National Academy of Sciences to secure the services of eminent members of appropriate professions in the examination of policy matters pertaining to the health of the public. The Institute acts under the responsibility given to the National Academy of Sciences by its congressional charter to be an adviser to the federal government and, upon its own initiative, to identify issues of medical care, research, and education. Dr. Harvey V. Fineberg is president of the Institute of Medicine.

The **National Research Council** was organized by the National Academy of Sciences in 1916 to associate the broad community of science and technology with the Academy's purposes of furthering knowledge and advising the federal government. Functioning in accordance with general policies determined by the Academy, the Council has become the principal operating agency of both the National Academy of Sciences and the National Academy of Engineering in providing services to the government, the public, and the scientific and engineering communities. The Council is administered jointly by both Academies and the Institute of Medicine. Dr. Ralph J. Cicerone and Dr. Charles M. Vest are chair and vice chair, respectively, of the National Research Council.

www.national-academies.org

Other Reports of the Space Studies Board and the Aeronautics and Space Engineering Board

Opening New Frontiers in Space: Choices for the Next New Frontiers Announcement of Opportunity (SSB, 2008)
Space Science and the International Traffic in Arms Regulations: Summary of a Workshop (SSB, 2008)
Workshop Series on Issues in Space Science and Technology: Summary of Space and Earth Science Issues from the Workshop on U.S. Civil Space Policy (SSB, 2008)

Assessment of the NASA Astrobiology Institute (SSB, 2007)
An Astrobiology Strategy for the Exploration of Mars (SSB with the Board on Life Sciences [BLS], 2007)
Building a Better NASA Workforce: Meeting the Workforce Needs for the National Vision for Space Exploration (SSB with the Aeronautics and Space Engineering Board [ASEB], 2007)
Decadal Science Strategy Surveys: Report of a Workshop (SSB, 2007)
Earth Science and Applications from Space: National Imperatives for the Next Decade and Beyond (SSB, 2007)
Exploring Organic Environments in the Solar System (SSB with the Board on Chemical Sciences and Technology, 2007)
Grading NASA's Solar System Exploration Program: A Midterm Review (SSB, 2007)
The Limits of Organic Life in Planetary Systems (SSB with BLS, 2007)
NASA's Beyond Einstein Program: An Architecture for Implementation (SSB with the Board on Physics and Astronomy [BPA], 2007)
Options to Ensure the Climate Record from the NPOESS and GOES-R Spacecraft: A Workshop Report (SSB, 2007)
A Performance Assessment of NASA's Astrophysics Program (SSB with BPA, 2007)
Portals to the Universe: The NASA Astronomy Science Centers (SSB, 2007)
The Scientific Context for Exploration of the Moon (SSB, 2007)

An Assessment of Balance in NASA's Science Programs (SSB, 2006)
Assessment of NASA's Mars Architecture 2007-2016 (SSB, 2006)
Assessment of Planetary Protection Requirements for Venus Missions: Letter Report (SSB, 2006)
Decadal Survey of Civil Aeronautics: Foundation for the Future (ASEB, 2006)
Distributed Arrays of Small Instruments for Solar-Terrestrial Research: Report of a Workshop (SSB, 2006)
Issues Affecting the Future of the U.S. Space Science and Engineering Workforce: Interim Report (SSB with ASEB, 2006)
Review of NASA's 2006 Draft Science Plan: Letter Report (SSB, 2006)
Review of the Space Communications Program of NASA's Space Operations Mission Directorate (ASEB, 2006)
The Scientific Context for Exploration of the Moon—Interim Report (SSB, 2006)
Space Radiation Hazards and the Vision for Space Exploration (SSB, 2006)

Limited copies of these reports are available free of charge from

Space Studies Board
National Research Council
The Keck Center of the National Academies
500 Fifth Street, N.W., Washington, DC 20001
(202) 334-3477/ssb@nas.edu
www.nationalacademies.org/ssb/ssb.html

NOTE: These reports are listed according to the year of approval for release, which in some cases precedes the year of publication.

PLANNING COMMITTEE FOR WORKSHOP ON U.S. CIVIL SPACE POLICY

LENNARD A. FISK, University of Michigan, *Chair*
CHARLES L. BENNETT, Johns Hopkins University
RAYMOND S. COLLADAY, Lockheed Martin Corporation (retired)
BERRIEN MOORE III, University of New Hampshire
GEORGE A. PAULIKAS, Aerospace Corporation (retired)
WARREN M. WASHINGTON, National Center for Atmospheric Research
A. THOMAS YOUNG, Lockheed Martin Corporation (retired)

RAPPORTEUR

MOLLY K. MACAULEY, Resources for the Future, Inc.

STAFF

JOSEPH K. ALEXANDER, Senior Program Officer, Space Studies Board
KERRIE SMITH, Program Officer, Aeronautics and Space Engineering Board
CARMELA J. CHAMBERLAIN, Program Associate, Space Studies Board
CATHERINE A. GRUBER, Assistant Editor, Space Studies Board
SANDRA WILSON, Program Assistant, Aeronautics and Space Engineering Board

SPACE STUDIES BOARD

LENNARD A. FISK, University of Michigan, *Chair*
A. THOMAS YOUNG, Lockheed Martin Corporation (retired), *Vice Chair*
SPIRO K. ANTIOCHOS, Naval Research Laboratory
DANIEL N. BAKER, University of Colorado
STEVEN J. BATTEL, Battel Engineering
CHARLES L. BENNETT, Johns Hopkins University
ELIZABETH R. CANTWELL, Los Alamos National Laboratory
ALAN DRESSLER, The Observatories of the Carnegie Institution
JACK D. FELLOWS, University Corporation for Atmospheric Research
FIONA A. HARRISON, California Institute of Technology
TAMARA E. JERNIGAN, Lawrence Livermore National Laboratory
KLAUS KEIL, University of Hawaii
MOLLY K. MACAULEY, Resources for the Future
BERRIEN MOORE III, University of New Hampshire
KENNETH H. NEALSON, University of Southern California
JAMES PAWELCZYK, Pennsylvania State University
SOROOSH SOROOSHIAN, University of California, Irvine
RICHARD H. TRULY, National Renewable Energy Laboratory (retired)
JOAN VERNIKOS, Thirdage LLC
JOSEPH F. VEVERKA, Cornell University
WARREN M. WASHINGTON, National Center for Atmospheric Research
CHARLES E. WOODWARD, University of Minnesota
GARY P. ZANK, University of California, Riverside

MARCIA S. SMITH, Director

AERONAUTICS AND SPACE ENGINEERING BOARD

RAYMOND S. COLLADAY, Lockheed Martin Astronautics (retired), *Chair*
CHARLES F. BOLDEN, JR., Jack and Panther, LLC
ANTHONY J. BRODERICK, Aviation Safety Consultant
AMY L. BUHRIG, Boeing Commercial Airplanes
PIERRE CHAO, Center for Strategic and International Studies
INDERJIT CHOPRA, University of Maryland
ROBERT L. CRIPPEN, Thiokol Propulsion (retired)
DAVID GOLDSTON, Princeton University
JOHN HANSMAN, Massachusetts Institute of Technology
PRESTON HENNE, Gulfstream Aerospace Corporation
JOHN M. KLINEBERG, Space Systems/Loral (retired)
RICHARD KOHRS, Independent Consultant
IVETT LEYVA, Air Force Research Laboratory, Edwards Air Force Base
EDMOND SOLIDAY, United Airlines (retired)

MARCIA S. SMITH, Director

Preface

In November 2003, the National Research Council's (NRC's) Space Studies Board (SSB), in collaboration with the NRC's Aeronautics and Space Engineering Board (ASEB), organized a workshop to encourage a continuing broad national discussion about the future direction of the U.S. civil space program.[1] The workshop was intended to explore aspects of the question, What should be the principal purposes, goals, and priorities of the U.S. civil space program? Participants observed attributes of NASA's science programs that were missing in the human exploration program and saw the opportunity to apply lessons learned from the comparison for the improvement of the human spaceflight program. A workshop report, *Issues and Opportunities Regarding the U.S. Space Program: A Summary Report of a Workshop on National Space Policy*,[2] was released on January 14, 2004.

Also on January 14, 2004, President George W. Bush announced a new national Vision for Space Exploration (the Vision); its fundamental goal was "to advance U.S. scientific, security, and economic interests through a robust space exploration program" that would involve human and robotic exploration of space, including sending humans back to the Moon and later to Mars.[3] The Vision had several cornerstones, including retiring the space shuttle by 2010, completing the International Space Station, and establishing a broad goal for human exploration of the Moon and, eventually, Mars. Subsequently, the June 2004 report of the President's Commission on Implementation of United States Space Exploration Policy[4] (known as the Aldridge Commission) articulated a balanced program for human and robotic space exploration and science. The NASA Authorization Act of 2005[5] demonstrated that Congress supports the Vision as part of a balanced program that includes science and aeronautics.

Two additional NRC reports, *Science in NASA's Vision for Space Exploration*[6] and *An Assessment of Balance in NASA's Science Programs*,[7] were prepared in response to congressional interest in relationships between the Vision and NASA's science programs. The one overarching finding of the latter report was that "NASA is being asked to accomplish too much with too little," and that "[t]he agency does not have the necessary resources to carry out the tasks of completing the International Space

---

[1] Participants at the 2003 workshop considered *civil space* to include all of NASA's human and robotic space programs; NOAA's meteorological and environmental satellite programs; the activities of commercial entities in support of the space programs of NASA, NOAA, and other civilian agencies; and commercial space activities. Military and national security reconnaissance space programs were not included under the rubric of civil space. Participants in the 2007 workshop took the same approach and also considered emerging entrepreneurial efforts such as space tourism to be part of civil commercial space.

[2] National Research Council, *Issues and Opportunities Regarding the U.S. Space Program: A Summary Report of a Workshop on National Space Policy*, The National Academies Press, Washington, D.C., 2004.

[3] National Aeronautics and Space Administration, *The Vision for Space Exploration*, NP-2004-01-334-HQ, NASA, Washington, D.C., 2004, pp. iii.

[4] President's Commission on Implementation of United States Space Exploration Policy, *A Journey to Inspire, Innovate and Discover* (also known as the Aldridge Commission report), June 2004, available at http://govinfo.library.unt.edu/moontomars/docs/M2MReportScreenFinal.pdf.

[5] The National Aeronautics and Space Administration Authorization Act of 2005, Public Law 109-155, 109th Congress, U.S. Government Printing Office, Washington, D.C., 2005.

[6] National Research Council, *Science in NASA's Vision for Space Exploration*, The National Academies Press, Washington, D.C., 2005.

[7] National Research Council, *An Assessment of Balance in NASA's Science Programs*, The National Academies Press, Washington, D.C., 2006.

Station, returning humans to the Moon, maintaining vigorous space and Earth science and microgravity life and physical sciences programs, and sustaining capabilities in aeronautical research."[8]

The problems of reconciling expectations for total program content and total program resources, sustaining support and momentum for human space exploration, and optimizing international cooperation and competition in space exploration have posed perennial challenges for policy makers, and they remain crucial today. Consequently, the NRC formed an ad hoc committee under the auspices of the SSB and the ASEB to organize a second public workshop to encourage national discussion about future directions of the U.S. civil space program. (See Appendix A for the statement of task.) The workshop, which was held on November 29-30, 2007, employed invited talks, panel discussions, and general discussions for reviewing developments that have occurred since the two boards held the 2003 workshop. See Appendix B for the workshop agenda. Approximately 60 workshop participants, whose expertise spanned the fields of human spaceflight, space science, commercial space, science and technology policy, economics, international relations, and the media, (see Appendix C) revisited aspects of the question, What are the principal purposes, goals, and priorities of U.S. civil space?, and they explored the following ancillary topics:

- Key changes and developments since 2003;
- How space exploration fits in a broader national and international context;
- Sustainability factors, including affordability, public interest, and political will;
- Definitions, metrics, and decision criteria for program portfolio mix and balance;
- Roles of government in Earth observations from space; and
- Requirements and gaps in capabilities and infrastructure.

The goal of the workshop was not to develop definitive answers to any of these questions but to air a range of views and perspectives that would serve to inform subsequent broader discussion of such questions by policy makers and the public.

This report presents a summary of the discussions at the November 2007 workshop. It is not intended to represent a consensus of the views of the workshop participants but to capture highlights of the discussions and to note major themes that emerged. In contrast, *Workshop Series on Issues in Space Science and Technology: Summary of Space and Earth Science Issues from the Workshop on U.S. Civil Space Policy*,[9] released in February 2008 as the first in a series of SSB workshop reports on issues in space science and technology, provided a brief synopsis that was limited to issues raised at the workshop that were particularly relevant to space and Earth science.

---

[8] National Research Council, *An Assessment of Balance in NASA's Science Programs*, The National Academies Press, Washington, D.C., 2006, p. 2.

[9] National Research Council, *Workshop Series on Issues in Space Science and Technology: Summary of Space and Earth Science Issues from the Workshop on U.S. Civil Space Policy*, The National Academies Press, Washington, D.C., 2008.

Acknowledgment of Reviewers

This report has been reviewed in draft form by individuals chosen for their diverse perspectives and technical expertise, in accordance with procedures approved by the National Research Council's Report Review Committee. The purpose of the independent review is to provide candid and critical comments that will assist the institution in making its published report as sound as possible and to ensure that the report meets institutional standards for objectivity, evidence, and responsiveness to the study charge. The review comments and draft manuscript remain confidential to protect the integrity of the deliberative process. We wish to thank the following individuals for their review of this report:

Alexander H. Flax, Independent Consultant,
George M. Hornberger, University of Virginia,
Joan Johnson-Freese, Naval War College,
Charles F. Kennel, Scripps Institution of Oceanography, University of California, San Diego, and
A. Thomas Young, Lockheed Martin Corporation (retired).

Although the reviewers listed above have provided many constructive comments and suggestions, they were not asked to endorse the statements presented in the report, nor did they see the final draft of the report before its release. The review of the report was overseen by W. Carl Lineberger, University of Colorado at Boulder. Appointed by the National Research Council, he was responsible for making certain that an independent examination of this report was carried out in accordance with institutional procedures and that all review comments were carefully considered. Responsibility for the final content of the report rests entirely with the authors and the institution.

Contents

SUMMARY ... 1

1 BACKGROUND ... 6
   2003 Space Policy Workshop, 6
   2004 Vision for Space Exploration, 7
   NASA Authorization Act of 2005, 8
   2006 National Space Policy, 9

2 ASSESSMENT OF THE CURRENT SITUATION ... 11
   Robustness, 11
   International Context, 13
   Public Interest and Support, 14

3 STRATEGIC ISSUES AND OPTIONS FOR SOLUTIONS ... 15
   Sustainability Factors, 15
      Realism About Resources, 15
      Leadership, 16
      Relevance and Value, 17
      Balance, 18
   Earth Observations, 19
   Capabilities and Infrastructure, 20

4 EPILOGUE ... 21
   Communicating about Space Exploration, 21
   International Competition, Cooperation, and Leadership, 21
   Ensuring Robustness Through New Approaches and Attitudes, 22

APPENDIXES

A Statement of Task ... 25
B Workshop Agenda ... 26
C Workshop Participants ... 28

Summary

What are the principal purposes, goals, and priorities of the U.S. civil space program?[1] This question was the focus of the workshop on civil space policy held November 29-30, 2007, by the Space Studies Board (SSB) and the Aeronautics and Space Engineering Board (ASEB) of the National Research Council (NRC). In addressing this question, invited speakers and panelists and the general discussion from this public workshop explored a series of topics, including the following:

- Key changes and developments in the U.S. civil space program since the new national Vision for Space Exploration[2] (the Vision) was articulated by the executive branch in 2004;
- The fit of space exploration within a broader national and international context;
- Affordability, public interest, and political will to sustain the civil space program;
- Definitions, metrics, and decision criteria for the mix and balance of activities within the program portfolio;
- Roles of government in Earth observations from space; and
- Gaps in capabilities and infrastructure to support the program.

The workshop organizers acknowledged the long-standing problem of reconciling expectations of civil space program accomplishments during the coming decades with the limited public resources available to support these activities. The goal of the workshop was neither to develop definitive solutions nor to reach consensus. Rather, the purpose was to air a range of views and perspectives that would serve to inform broader discussion of such questions by policy makers and the public. This document summarizes the opinions expressed by individual workshop participants and does not necessarily reflect the consensus views of these participants, the SSB, or the workshop planning committee.

By way of background, the SSB and the ASEB had convened a similar workshop in 2003 in the wake of the space shuttle Columbia tragedy and the findings of the Columbia Accident Investigation Board. Since the issuance of the report on the 2003 workshop, *Issues and Opportunities Regarding the U.S. Space Program: A Summary Report of a Workshop on National Space Policy*,[3] additional developments have taken place to redirect many elements of the civil space program. The Vision for Space Exploration set forth by the executive branch in 2004, the National Aeronautics and Space Administration (NASA) Authorization Act of 2005,[4] and the national space policy presidential directive issued in 2006 have all served to redirect the program. The Vision sets forth a long-term robotic and human exploration program; the NASA Authorization Act of 2005 endorses the Vision and directs the

---

[1] Participants at the 2003 workshop considered *civil space* to include all of NASA's human and robotic space programs; NOAA's meteorological and environmental satellite programs; the activities of commercial entities in support of the space programs of NASA, NOAA, and other civilian agencies; and commercial space activities. Military and national security reconnaissance space programs were not included under the rubric of civil space. Participants in the 2007 workshop took the same approach and also considered emerging entrepreneurial efforts such as space tourism to be part of civil commercial space.

[2] National Aeronautics and Space Administration, *The Vision for Space Exploration*, NP-2004-01-334-HQ, NASA, Washington, D.C., 2004.

[3] National Research Council, *Issues and Opportunities Regarding the U.S. Space Program: A Summary Report of a Workshop on National Space Policy*, The National Academies Press, Washington, D.C., 2004.

[4] The National Aeronautics and Space Administration Authorization Act of 2005, Public Law 109-155, 109th Congress, U.S. Government Printing Office, Washington, D.C., 2005.

program in several areas with respect to policy, management, and accountability and oversight; and the 2006 presidential directive establishes goals related to U.S. space leadership and the governance of space operations in and through space.

ROBUSTNESS OF THE CIVIL SPACE PROGRAM

The workshop summarized here thus builds on discussion from the 2003 workshop in light of these developments. A natural starting point was an assessment of the new directions for the U.S. civil space program: How robust or resilient are these new directions to changes in resources available to support the program? How relevant is the program in what many workshop participants see as a rapidly changing international context? Is there public appeal in terms of willingness to embrace the program? Many participants expressed the view that the Vision had not progressed as originally outlined nor as many had expected, due in large part to the failure of the administration and the Congress to seek the required resources. A prominent concern among participants was that although the Vision was to be "pay as you go," shortfalls in the NASA budget had led the agency to reallocate resources toward pursuit of the Vision and away from other activities such as space and Earth science. Speakers argued that continued operational costs of the International Space Station, delayed phaseout of the space shuttle, costs of near-term development of the next-generation space transportation system, and unbudgeted operational costs will all make the Vision increasingly unaffordable. Other participants acknowledged that some of the problems with robustness and program balance are of the space community's own making, in that in many activities, project cost estimates had been unrealistic and subject to significant cost growth. Participants from within and outside the scientific community voiced agreement that the community will need to demonstrate leadership and share responsibility with NASA in controlling science program costs. Speakers expressed concern that NASA's program suffers from a lack of resources, budget realism, and budget stability, thereby making the Vision unaffordable and unsustainable.

The recent report that focused on the space and Earth science issues at this workshop summarized the mood at the workshop as follows:[5]

> Overall, as noted by the participants themselves, the tone of the workshop was surprisingly sober, with frequent expressions of discouragement, disappointment, and apprehension about the future of the U.S. civil space program. During the one and one-half days of discussion, an oft-repeated statement by workshop participants was that the goals of the U.S. civil space program are completely mismatched with the resources provided to accomplish them.

INTERNATIONAL CONTEXT

In contrast with the 2003 workshop at which international developments were mentioned but did not play a pivotal role in discussion, international collaboration and competition were prominent topics at the 2007 workshop. Speakers summarized their understanding of the capabilities and ambitions of new national space programs in China and India, cited the forming of multinational alliances that exclude the United States or Europe, and pointed out some consequences of the U.S. International Traffic in Arms Regulations (ITAR) as examples of new challenges in balancing cooperation and competition in the U.S. civil space program. For example, speakers questioned whether a goal of cooperation conflicts with the objective in the Vision to support international participation "to further U.S. scientific, security, and

---

[5] National Research Council, *Workshop Series on Issues in Space Science and Technology: Summary of Space and Earth Science Issues from the Workshop on U.S. Civil Space Policy,* The National Academies Press, Washington, D.C., 2008, p. 2.

economic interests."[6] Some participants suggested that international cooperation could provide a means to share costs, thereby augmenting resources available for the space program, but others noted that collaboration does not always result in reduced costs, particularly if partner roles and responsibilities are unclear. Participants also discussed at length the emergence of China as a major player in space and whether China presents a threat, in which case cooperation may be difficult or even out of the question, or an opportunity for engagement and cooperation, in which case space could gain a new strategic purpose as a vehicle for such cooperation. In any case, discussion highlighted that a decision about how to engage China will not be based solely on space policy, but will depend on much larger geopolitical considerations.

PUBLIC INTEREST AND SUPPORT

In assessing contemporary public interest in and support for space activities, some participants commented that programs such as the Hubble Space Telescope and the Mars rovers are popular and have a "wow factor"; other speakers suggested that as long as the NASA budget is not too large, a "wow factor" in space accomplishments becomes less important. Others noted some survey-based evidence[7] that the greatest degree of enthusiasm for human space exploration rests with the Apollo generation (the 45- to 64-year-old age group), with much less support from the generation of youngest voters—the 18- to 24-year-old age group.

SUSTAINABILITY, RESOURCES, LEADERSHIP, RELEVANCE, AND BALANCE

Subsequent discussion turned to identifying problems in more detail, specifically to addressing a lack of resources, leadership challenges, the relevance and value of the space program, and balance among activities within the program. Speakers cited both internal and external factors that can affect resource requirements. Internal factors include project delays, inadequate contingency funds, pressures for "full employment" at NASA centers, and defensive behavior by program managers and others when resources are scarce. External influences include competition from China and India, the emergence of climate and energy as major global issues, and likely continued federal budget deficits. Another concern was potential congressional opposition to U.S. reliance on Russia during an extended launch hiatus after the retirement of the space shuttle.

The question of leadership figured prominently in workshop discussions. Some participants argued that strong leadership at senior levels of NASA and the government is essential for the success of the space program. In this context, some speakers viewed with considerable urgency the desirability of senior leaders facing up to what was repeatedly described as a program that cannot be executed within the allotted budget. Speakers also reiterated the responsibility of the space community to establish sound cost estimates and to execute programs within realistic budgets.

Why should I care?—suggested by a participant as an appropriate question to be posed by candidates for major national office—served to focus in-depth discussion about a rationale for the civil space program. There were considerable differences in opinion, ranging from historically offered reasons (science, national security, commercial activities, a sense of human destiny and exploration, and national prestige and geopolitics) to a focus on the geopolitical contributions of the space program as perhaps one of the most compelling current-day rationales. But there was less than full agreement as to whether

---

[6] National Aeronautics and Space Administration, *The Vision for Space Exploration*, NP-2004-01-334-HQ, NASA, Washington, D.C., 2004, p. iii.

[7] M.L. Dittmar, *Engaging the 18-25 Generation: Educational Outreach, Interactive Technologies, and Space*, Dittmar Associates, Inc., available at http://www.dittmar-associates.com/Publications/Engaging%20the%2018-25%20Generation%20Update~web.pdf.

geopolitics meant cooperation or competition as a motivation for space activities. Discussion also addressed but did not reach agreement on whether, and if so to what extent, the civil space program needs to demonstrate practical benefits and value, a "wow" factor, or some mix of both.

Balancing the pursuit of science, human space exploration, aeronautics, and other dimensions of space activities was also a concern among participants. Some speakers cautioned against characterizing the problem as "humans versus robots"; others urged that the focus should be on identifying and exploiting synergies among different parts of NASA, among NASA and other agencies and countries, and between NASA and the private sector. Participants also suggested that assessing balance requires recognition that different constituencies have different objectives—for example, the scientific community measures much of its success in terms of progress toward goals such as those articulated in decadal surveys, whereas the aeronautics community measures progress in terms of responding to commercial and military air transport requirements.

EARTH OBSERVING PROGRAMS

Workshop discussion also addressed the role of Earth observations. Speakers emphasized that Earth observations necessarily assume even greater importance given evidence of possibly significant changes in climate. But they remained troubled by problems stemming from reorganization of responsibility for and funding of the National Polar-Orbiting Operational Environmental Satellite System (NPOESS) and the reduced capability of NPOESS in facilitating necessary climate-related measurements. Discussion also addressed the persistent difficulty between NASA and the National Oceanic and Atmospheric Administration (NOAA) in the "handoff" from use for research purposes to operational use of Earth science infrastructure and information. Speakers argued that differences in these agencies—ranging from culture to objectives—become even sharper when their budgets are declining.

CAPABILITIES AND INFRASTRUCTURE

Additional workshop discussion included optimistic comments about future capabilities and infrastructure to support the civil space program if national priorities can be well articulated and sufficient resources made available. For example, both traditional and new companies in aerospace can bring creativity and talent to problem solving when requirements are made clear. Speakers described experiences with bright university students interested in aerospace careers provided students sense that they can have an impact. Speakers further urged that NASA and universities build more effective partnerships to encourage talent and that ITAR restrictions limiting access to good students be remedied. Some participants mentioned institutions where turnover rates among aerospace professionals are very low, even at the present time. Discussion also addressed the attraction of many young people to space activities using contemporary media that create a virtual presence.

CONCLUDING THEMES

The workshop concluded with the consolidation of discussion topics, which fell into three broad categories: communicating about space exploration; international competition, cooperation, and leadership; and ensuring robustness through new approaches and attitudes. One idea for avoiding the impending programmatic "train wreck" to which many participants referred during the workshop was to "slow down the train" by deferring the first human mission to the Moon; extending the use of the

International Space Station in support of research and development for later human exploration; establishing a telepresence on the Moon; creating an environment of institutional stability in NASA's program elements; building globally inclusive working groups on direct missions to Mars, global change, and space science; and defining real, meaningful jobs for humans in space.

1

Background

Several notable developments in U.S. civil space policy in the years between the National Research Council's (NRC's) 2003 space policy workshop and its 2007 workshop are summarized in this chapter. These developments were captured in key documents made available to workshop participants as background information that participants referred to during the workshop discussions.

2003 SPACE POLICY WORKSHOP

In November 2003, the NRC's Space Studies Board and Aeronautics and Space Engineering Board organized a workshop to air perspectives on the question, What should be the principal purposes, goals, and priorities of U.S. civil space policy? The timing of that workshop coincided with a new attention to the long-term direction of the U.S. civil space program in the wake of the space shuttle Columbia tragedy and the report of the Columbia Accident Investigation Board (CAIB).[1] Seven broad themes emerged from the 2003 workshop:[2]

1. *Successful space and Earth science programs*—Many of the 2003 workshop participants accepted that U.S. space and Earth science programs were currently productive and progressing steadily. Much of the success of NASA's science programs was attributed to having clear long-range goals; strategies framed by scientists and periodically reassessed by the science community; and a series of individual steps that accumulate successes, help measure progress, and sustain momentum for the program.

2. *A clear goal for human spaceflight*—Many 2003 workshop participants echoed the CAIB's conclusion that a lack of an agreed vision for the human spaceflight program had had a negative impact on the health of that program in NASA. A bold goal could enable breaking out of programmatic drift, providing a transcendent purpose for the risk of human endeavors in space and the opportunity for leadership if the United States would openly invite others to participate in setting and steadily pursuing a shared long-range goal.

3. *Exploration as the goal for human spaceflight*—Many 2003 workshop participants emphasized two fundamental reasons to send humans to space:

—Exploration can and does add to the acquisition of new knowledge, that is, knowledge of space as a place for human activity and knowledge of the solar system, including Earth, from the vantage point of space.

—Exploration is a basic human desire, a general impulse of human nature.

---

[1] Columbia Accident Investigation Board, *Columbia Accident Investigation Board Report: Volume 1*, available at http://caib.nasa.gov/, August 2003.

[2] National Research Council, *Workshop Series on Issues in Space Science and Technology: Issues and Opportunities Regarding the U.S. Space Program: A Summary Report of a Workshop on National Space Policy*, The National Academies Press, Washington, D.C., 2004.

4. *Exploration as a long-term endeavor to be accomplished by means of a series of small steps*—Many participants argued that having a clear, agreed-on, long-term goal, such as the human exploration of Mars, is essential for the future success of the human spaceflight program, but that it is premature to set a firm date for or cost of that goal. What is possible is a first assessment of what has to be accomplished, and the identification of intermediate, subsidiary goals that can be met in a series of smaller steps and would evolve at a pace that reflects a meaningful rate of learning.

5. *Synergy superseding the humans-versus-robots dichotomy*—The ultimate achievement of a long-term goal for human exploration, numerous participants argued, should be to best employ both human and robotic assets and to have the space program move beyond complementarity and toward a synergy between robots and humans. Whatever the destination and whatever the specific means chosen, many participants stated that being guided by a principle of synergy between robots and humans provides the opportunity to explore the solar system in the most optimal manner.

6. *The long-term goal driving all implementation decisions*—Participants in the 2003 workshop appeared to view the following activities as essential elements along the path to a goal for human exploration:

—The continued robotic exploration of our solar system followed by the development of capable human-machine interfaces and teleoperators,

—Research on the International Space Station focused on addressing the questions posed by human exploration away from low Earth orbit, and

—Development of a space transportation system to replace the shuttle, all directed toward facilitating the eventual human exploration of some destination beyond low Earth orbit.

7. *Institutional concerns*—The first six themes represented crosscutting concepts relevant to the nation's future approach to civil space. The seventh theme collected the views offered by the 2003 workshop participants on needs and opportunities for successful implementation of future space policy in three areas:

—Cross-institutional or cross-sector activities—for example, engaging in joint technology development, taking advantage of synergies, and improving planning and development—all of which were seen as dependent on the availability of a skilled industrial base;

—NASA as the primary executive branch agency responsible for implementing space policy; and

—The scientific community, one of NASA's key constituents.

2004 VISION FOR SPACE EXPLORATION

On January 14, 2004, President George W. Bush announced the new national Vision for Space Exploration, whose fundamental goal was "to advance U.S. scientific, security, and economic interests through a robust space exploration program" by means of "an integrated, long-term robotic and human exploration program with measurable milestones and executed on the basis of available resources, accumulated experience, and technology readiness."[3] The Vision called for a set of key activities in four areas, as follows:

- *Low Earth orbit.* Use of the space shuttle would be focused on completing the assembly of the International Space Station (ISS), and then the shuttle would be retired by the end of the decade (i.e., 2010). Use of the ISS would be focused on supporting exploration goals "in a manner consistent with U.S. obligations" between the United States and other partners.
- *Beyond low Earth orbit.* Lunar exploration activities would be designed to enable the exploration of Mars and beyond by means of robotic missions starting in 2008 and human missions no

---

[3] National Aeronautics and Space Administration, *The Vision for Space Exploration*, NP-2004-01-334-HQ, NASA, Washington, D.C., 2004, pp. iii-iv.

later than 2020. Mars robotic exploration would continue, leading to later human missions after successful demonstration on the Moon. Robotic exploration would continue across the solar system and would be complemented by telescopic searches around other stars. In addition, there would be demonstrations of key capabilities to support ambitious human and robotic exploration.

- *Space transportation capabilities.* A program (subsequently named Project Constellation) would support the design and development of a new crew exploration vehicle for missions beyond low Earth orbit (the Orion spacecraft) and would thus separate crew transportation vehicle (Ares, the cargo-launch component).
- *International and commercial participation.* The United States would pursue opportunities for international participation to support U.S. space exploration goals and pursue commercial opportunities for providing transportation and other services supporting the ISS and exploration beyond low Earth orbit.

To provide the resources to accomplish the Vision, the administration proposed budget increases of 5 percent per year for 3 years (fiscal year [FY] 2005 through FY 2007) and then 1 percent increases in the following 2 years.[4] The budget strategy relied on holding down growth in programs that did not support the Vision, freeing billions of dollars in the decade beyond 2010 by retiring the shuttle, and finding innovative approaches to reduce the costs of space operations.

The president's announcement also called for the formation of the President's Commission on Implementation of United States Space Exploration Policy, which would be charged to make recommendations on implementing the Vision. The commission's report, which was released in June 2004,[5] provided findings and recommendations about NASA management, development of enabling technologies, roles of the private sector and international participants, scientific research as a part of exploration, and opportunities for education and public engagement. In order to manage the exploration programs within the resources that were expected to be available, the commission recommended a "go as you can pay" approach that would allow specific goal milestones to be adjusted depending on what could be afforded along the way.

NASA AUTHORIZATION ACT OF 2005

In the NASA Authorization Act of 2005 (Public Law 109-155),[6] which was enacted on December 30, 2005, Congress gave NASA program responsibilities for FY 2006 through FY 2008, and it authorized appropriations for FY 2007 and FY 2008. The act endorsed the Vision, and it provided guidance and direction in several areas with respect to policy, program management, and accountability and oversight. The Joint Explanatory Statement of the House-Senate conference committee for the bill[7] indicated as follows:

> The conferees believe that the Conference Report provides a strong legislative foundation for the pursuit of the nation's continued exploration of space in a manner that both preserves the

---

[4] The FY 2005 NASA budget request called for a total budget of $16.2 billion in FY 2005, rising to $17.8 billion in FY 2007, and reaching $18.0 billion in FY 2009. The actual totals appropriated by Congress were $16.2 billion in FY 2005 and $16.3 billion in FY 2007; the administration's request for FY 2009 in $17.6 billion.

[5] President's Commission on Implementation of United States Space Exploration Policy, *A Journey to Inspire, Innovate and Discover* (also known as the Aldridge Commission report), June 2004, available at http://govinfo.library.unt.edu/moontomars/docs/M2MReportScreenFinal.pdf.

[6] The National Aeronautics and Space Administration Authorization Act of 2005, Public Law 109-155, 109th Congress, U.S. Government Printing Office, Washington, D.C., 2005.

[7] Joint Explanatory Statement of the Committee of Conference, Conference Report on S. 1281, National Aeronautics and Space Administration Authorization Act of 2005, U.S. House of Representatives, December 16, 2005, *Congressional Record*, Volume 151, p. H12028.

important legacy of accomplishments in science, aeronautics and human space flight and provides NASA with the authority to move its new program of exploration forward.

The statement also provided a set of priorities that included the following:[8]

- A continued strong and diverse array of programs in the areas of space science, earth science and education is essential;
- [A] need to provide the smoothest possible transition between the eventual retirement of the space shuttle and the development of the new Crew Exploration Vehicle (CEV) and Crew Launch Vehicle (CLV);
- [The] research potential of the ISS beyond its contribution to long-duration human spaceflight in support of the Vision for Space Exploration; and
- A national aeronautics research policy to guide future investments in this important segment of NASA's mission . . . to ensure the vitality of aeronautics research within the framework of a clear set of national policy objectives.

The NASA Authorization Act of 2005 specified NASA's general responsibilities as follows:[9]

The Administrator shall ensure that NASA carries out a balanced set of programs that shall include, at a minimum, programs in—

(A) Human space flight, in accordance with [bill language setting milestones for elements of the Vision, including the Crew Exploration Vehicle as close to 2010 as possible and returning Americans to the Moon no later than 2020];
(B) Aeronautics research and development; and
(C) Scientific research, which shall include, at a minimum—
   (i) Robotic missions to study the Moon and other planets and their moons, and to deepen understanding of astronomy, astrophysics, and other areas of science that can be productively studied from space;
   (ii) Earth science research and research on the Sun-Earth connection through the development and operation of research satellites and other means;
   (iii) Support of university research in space science, earth science, and microgravity science; and
   (iv) Research on microgravity, including research that is not directly related to human exploration.

2006 NATIONAL SPACE POLICY

In September 2006, President Bush authorized a new national space policy[10] that was intended to govern all U.S. national security and civil space activities. The policy reaffirmed the long-standing commitment to the peaceful exploration and use of space, including the use of space for "U.S. defense and intelligence-related activities in pursuit of national interests." It also did the following:

- Rejected any claims to sovereignty by any nation over outer space or celestial bodies,

---

[8] Joint Explanatory Statement of the Committee of Conference, Conference Report on S. 1281, National Aeronautics and Space Administration Authorization Act of 2005, U.S. House of Representatives, December 16, 2005, *Congressional Record*, Volume 151, p. H12028.

[9] The National Aeronautics and Space Administration Authorization Act of 2005, Public Law 109-155, 109th Congress, U.S. Government Printing Office, Washington, D.C., 2005.

[10] Office of Science and Technology Policy, *U.S. National Space Policy*, National Security Presidential Directive 49, unclassified version released on October 6, 2006, available at http://www.ostp.gov/galleries/default-file/Unclassified%20National%20Space%20Policy%20--%20FINAL.pdf, p. 1.

- Supported cooperation with other nations in the peaceful use of outer space,
- Considered space systems to have the rights of passage through and operations in space without interference,
- Asserted that the U.S. will preserve its rights, capabilities, and freedom of action in space and take those actions necessary to protect its space capabilities,
- Opposed the development of new legal regimes or other restrictions that seek to prohibit or limit U.S. access to or use of space, and
- Committed to encouraging and facilitating a growing and entrepreneurial U.S. commercial space sector.

The policy set forth seven fundamental goals, as follows:[11]

- Strengthen the nation's space leadership and ensure that space capabilities are available in time to further U.S. national security, homeland security, and foreign policy objectives;
- Enable unhindered U.S. operations in and through space to defend our interests there;
- Implement and sustain an innovative human and robotic exploration program with the objective of extending human presence across the solar system;
- Increase the benefits of civil exploration, scientific discovery, and environmental activities;
- Enable a dynamic, globally competitive domestic commercial space sector in order to promote innovation, strengthen U.S. leadership, and protect national, homeland, and economic security;
- Enable a robust science and technology base supporting national security, homeland security, and civil space activities; and
- Encourage international cooperation with foreign nations and/or consortia on space activities that are of mutual benefit and that further the peaceful exploration and use of space, as well as to advance national security, homeland security, and foreign policy objectives.

In addition to providing implementation guidelines for the secretary of defense and the director of national security for the national security space program, the policy provided civil space program guidance for NASA and the Department of Commerce as follows:[12]

> The United States shall increase the benefits of civil exploration, scientific discovery, and operational environmental monitoring activities. To that end, the Administrator, National Aeronautics and Space Administration shall: execute a sustained and affordable human and robotic program of space exploration and develop, acquire, and use civil space systems to advance fundamental scientific knowledge of our Earth system, solar system, and universe.
>
> The Secretary of Commerce, through the Administrator of the National Oceanic and Atmospheric Administration, shall in coordination with the Administrator, National Aeronautics and Space Administration, be responsible for operational civil environmental space-based remote sensing systems and management of the associated requirements and acquisition process.

---

[11] Office of Science and Technology Policy, *U.S. National Space Policy*, National Security Presidential Directive 49, unclassified version released on October 6, 2006, available at http://www.ostp.gov/galleries/default-file/Unclassified%20National%20Space%20Policy%20--%20FINAL.pdf, p. 2.

[12] Office of Science and Technology Policy, *U.S. National Space Policy*, National Security Presidential Directive 49, unclassified version released on October 6, 2006, available at http://www.ostp.gov/galleries/default-file/Unclassified%20National%20Space%20Policy%20--%20FINAL.pdf, p. 5.

2

Assessment of the Current Situation

The first two sessions of the 2007 workshop were intended to promote discussion of the current state of the U.S. civil space program and consideration of how civil space exploration fits in a larger national and international context. Each session began with remarks from panelists invited to address several framing questions for the session (see Appendix B), after which the session was opened for general discussion by all workshop participants.

The first session—on situational assessment—was moderated by Space Studies Board (SSB) member A. Thomas Young (Lockheed Martin Corporation, retired); the panel members were Bretton Alexander (X Prize Foundation), Fiona Harrison (California Institute of Technology), and James Zimmerman (International Space Services, Inc.). The session focused on identifying key developments and changes with respect to the U.S. civil space program since the 2003 SSB-Aeronautics and Space Engineering Board (ASEB) workshop. SSB member Charles Bennett (Johns Hopkins University) moderated the next session—on the national and international context for space; the panelists were journalist and author Guy Gugliotta, Joan Johnson-Freese (Naval War College), and Roger Launius (National Air and Space Museum).

Both sessions brought forth several common themes to which participants often returned for discussion and elaboration, both during the two sessions and later during the workshop. Throughout all the discussions, few speakers voiced opposition to several core views—namely, that the civil space program has been effective and important in the past in spite of its problems, that human space exploration is and will remain an important element of U.S. civil space policy, that space and Earth science and aeronautics are also essential central responsibilities of the civil space program, and that while each of these segments of the program has its own unique problems, human space exploration is now especially vulnerable.

Key aspects of the recurring themes that were highlighted during the discussions are summarized below.

ROBUSTNESS

Many participants expressed the view that the Vision for Space Exploration (the Vision) had not progressed as it was presented or as people in the space community had expected when the Vision was announced in 2004. Speakers argued that neither the administration nor Congress had sought the resources that would be required to accomplish the Vision. A significant consequence of this situation, people noted, is that resource shortfalls in budgets to support the development of new exploration systems integral to the Vision are having major disruptive impacts on other parts of NASA's programs. One speaker captured the spirit of the discussions by noting that the situation is "like a game of musical chairs, but we are more than just one chair short."

In analyzing the robustness of the program, speakers identified several concerns. First, there was concern that there was little substantial follow-up by the administration after the initial announcement of the Vision. One speaker contrasted the situation with that following President John F. Kennedy's announcement of the Apollo Program in which the president was explicit about the need to commit the necessary resources to the program and subsequently walled off the resources so that they would be protected.

Inadequate implementation was a second factor cited as having compromised the robustness of the Vision. One speaker noted that, in contrast to the presentation of the Vision as a new direction, there exists today a business-as-usual sense about a program that will be neither affordable nor sustainable. The idea of portraying the Vision as "Apollo on steroids" was cited as being particularly unsustainable. Initially it had also been expected that there would be significant public participation in crafting and executing the Vision, but these expectations have not been borne out. Another speaker commented that the Vision's "new direction" was also to have included a role for the commercial sector, but that too few opportunities had been provided to this sector. Additional discussion about the potential emergence of commercial space capabilities to support the Vision asked whether these capabilities could develop beyond serving only government (for example, commercial resupply services for the International Space Station) to serve private markets (such as space tourism). A speaker suggested that the "window" for new commercial space transportation companies focused on tourism was a limited window—perhaps 3 to 5 years. That speaker also commented that if the private spaceflight market succeeded, then "everything changes" in terms of the perception of the commercial sector's role in space. Several speakers were critical of the "go-as-you-can-pay" approach that was a premise for executing the Vision. They argued that such an approach is not realistic or feasible for carrying out complex technical tasks, because such efforts will lack the necessary flexibility to deal with technical developments and obstacles along the way. While the go-as-you-can-pay approach might be useful when initial decisions are being made, that is not the case afterward.

A third factor creating concern about robustness was that NASA is inadequately funded and that it is unlikely that political decision makers will relax the overall funding limits for NASA in the near future. That is, flat budgets are not likely to go away soon. Speakers argued that continued operational costs for the International Space Station, delayed phaseout of the space shuttle, costs of pressing near-term development of the next-generation space transportation system, and unbudgeted operational costs to achieve announced goals will all make the Vision unaffordable.

One consequence of this dilemma, several speakers noted, is that NASA's programs in space science, Earth science, and aeronautics are being affected in ways that will have serious long-term consequences. One speaker described NASA's science program as being "in retreat," citing recent program changes and management turbulence, the effects of which rapidly propagate across NASA and undermine intra-agency, interagency, and international planning and cooperation. Thus, while NASA is still reaping the scientific and public-image benefits from investments in science programs made 10 to 15 years ago, the speaker argued that the "free ride is about to end." Participants also acknowledged that some of the problems with the robustness and balance in the science program are of the program's own making, because of the impact of unrealistic past project cost estimates and significant cost growth. Participants from within and outside the scientific community voiced agreement that the scientific community will need to exert leadership and share responsibility with NASA to make tough decisions about controlling the science program costs.

Further discussion of these points was wide-ranging. There was reference to whether a "base realignment and closing" strategy could be used to trim or realign NASA centers, discussion of the difficulty of outsourcing external to NASA centers given consolidation in the aerospace industry (there are too few companies to which to outsource), and the question of whether another aspect of "robustness" might consider whether the nation and its space activities have become too "risk-averse" in willingness to try new technologies or more creative approaches.

In summing up participants' comments, a speaker noted that there are three essential elements of an endeavor—goals, a strategy to achieve the goals, and the means to implement the strategy—but that for NASA the means are lacking, thereby making the goals "a fantasy." Citing a 2006 report of the National Research Council which concluded that "NASA is being asked to accomplish too much with too little,"[1] speakers argued that NASA's program suffers from a lack of resources, budget realism, and budget stability, thereby making the Vision unsustainable and unachievable.

INTERNATIONAL CONTEXT

In contrast with the 2003 workshop, at which international considerations were mentioned but did not play a pivotal role, issues regarding international collaboration and competition were frequent discussion topics at the 2007 workshop. Several speakers noted that at least two factors inhibit international relationships in civil space programs. One relates to the Vision's statement of support for international participation "to further U.S. scientific, security, and economic interests"[2] and its emphasis on long-range goals at the expense of nearer-term opportunities for international cooperation, and to the fact that military space activity rather than civil space has received the attention of the current administration. Speakers noted that by failing to emphasize international cooperation in using the International Space Station for near-term enabling research in support of human exploration, an opportunity to engage international partners in early implementation of the Vision was lost. The other notable impediment mentioned was the implementation of the International Traffic in Arms Regulations (ITAR) and current U.S. government emphasis on military space activities, which were described as having a chilling impact on current and future cooperative relationships with foreign space partners.

Speakers also focused on the growing ambitions and capabilities of other nations' new national space programs, especially in China and India. Participants commented that there is an expanding level of capability in space activities across the world and that more nations are acquiring capabilities that make them strong collaborators and/or competitors with the United States. In addition, new arrangements are excluding the United States or Europe. China and India are forming their own respective alliances, such as cooperation between India and Israel in launch capability and between China and Brazil for Earth observations. Space-faring countries abroad are no longer considering the United States or Europe as default partners for collaboration.

Some participants also noted that international cooperation provides a means to share the costs of large, expensive programs and thereby to make them more affordable for individual countries. However, others noted that international cooperation as a means of reducing costs needs to be considered with care. First, international collaborative projects do not always result in reduced costs, especially when the separation of partner roles and responsibilities is blurred. Second, although international cooperation can make an expensive project affordable, it also can translate into yielding the country's international technological lead when other partners are brought in to the enterprise. Citing the Global Earth Observation System of Systems (GEOSS), a speaker asked if a "GEOSS-like" arrangement would be useful or workable for space science.

Finally, workshop participants devoted considerable attention to the emergence of China as a major player in space. Some speakers described China as presenting two alternative choices for relationships. If the United States sees China as a threat or as mainly a competitor, then cooperation may be difficult or even out of the question. Alternatively, the United States could seek engagement with China, in which case space exploration could gain a new strategic purpose as an element of engagement and cooperation. Speakers noted than even during the Cold War, when relationships with the Soviet Union were especially tense, there was continuing cooperation in space research. Several speakers argued that a decision about cooperation with China will not be a matter of "whether" but of "when and how." One speaker introduced the concept of the United States as a "benevolent hegemon." That is, there is an opportunity for the space program to become transformational as a means to exert U.S. leadership in

---

[1] National Research Council, *An Assessment of Balance in NASA's Science Programs*, The National Academies Press, Washington, D.C., 2006, p. 2.

[2] National Aeronautics and Space Administration, *The Vision for Space Exploration*, NP-2004-01-334-HQ, NASA, Washington, D.C., 2004, p. iii.

working with China for the betterment of the world. In any case, some speakers noted that a decision about how to engage China will not be based solely on space policy but will depend on much larger geopolitical considerations.

Several speakers posed questions such as these: What roles can international cooperation play in solutions to problems with the civil space program? Do we need a new paradigm for partnerships? Should international relationships be more central and more often on the critical path? Should the United States be more proactive? A frequent response was that international aspects cannot be ignored and that space activities offer a natural vehicle for establishing connectivity between nations, which in turn is an essential element of globalization and, therefore, a necessary priority of forward-looking nations.

PUBLIC INTEREST AND SUPPORT

During the session on national and international context and often in subsequent discussions, participants turned to an assessment of the degree of contemporary public interest and support for space exploration. Providing some historical perspective, speakers indicated that there was considerable public apathy during the Apollo program, with less than a majority of the public supporting sending people to the Moon. Instead, Apollo drew much of its strongest support from its links to political goals and priorities driven by the Cold War. Additional discussion centered on a longer-term problem: once we return to the Moon, it will "cost a lot to stay on the Moon," and once we get to Mars, it will "cost a lot to stay on Mars." In the case of Apollo, once we got to the Moon, we terminated the program.

One speaker described public support for space exploration as "a mile wide and an inch deep" and attributed some current public apathy to changes in public attitudes and expectations over the past decades. That is, people now have shorter attention spans and expect to have a more participative experience than is now offered by much of space exploration. This perspective was reinforced later in the workshop by citation of survey data that had been prepared for George Mason University,[3] indicating that today the greatest degree of enthusiasm for human space exploration rests with the 45- to 64-year-old age group (the Apollo generation) but that support is weak in the 18- to 24-year-old age group. In contrast, other speakers argued that people do care about space and that members of the lay public respond especially to the "wow factor"—that is, programs such as the Hubble Space Telescope and Mars exploration do capture public interest and support. However, a speaker commented that if the NASA budget is not viewed to be particularly large, the need to have a "wow" factor to sustain public support becomes less important. Speakers also noted that the level of excitement over space exploration appears to be much higher in countries abroad than inside the United States.

The workshop discussions drew out several competing views about what aspects of space exploration are most relevant and effective in engaging public interest. Participants discussed whether factors that were important in the past remain so today or whether new arguments and attributes will be more important as the country looks to the future. These ideas were explored in more detail in later sessions. (See Chapter 3.)

---

[3] M.L. Dittmar, *Engaging the 18-25 Generation: Educational Outreach, Interactive Technologies, and Space*, Dittmar Associates, Inc., available at http://www.dittmar-associates.com/Publications/Engaging%20the%2018-25%20Generation%20Update~web.pdf.

3

Strategic Issues and Options for Solutions

Following the pair of scene-setting sessions on the first day of the workshop, the second day provided participants with an opportunity to examine several strategic issues in more detail and to consider options for solutions to problems that were cited in earlier discussions. As was the case during prior sessions, there were a number of recurring themes, which this chapter briefly summarizes.

SUSTAINABILITY FACTORS

The session on sustainability explored two related themes: (1) long-term approaches for reconciling mismatches between expansive expectations for the civil space program and available resources and (2) factors that influence support from the public and from policy makers for long-term space exploration. Space Studies Board (SSB) member James Pawelczyk (Pennsylvania State University) moderated the session, which began with comments from panelists Paul Carliner (independent consultant and former congressional appropriations committee staff member), George Paulikas (The Aerospace Corporation, retired), Richard Truly (National Renewable Energy Laboratory, retired), and George Whitesides (National Space Society). The subsequent session on balance issues focused on principles for setting priorities for allocating responsibilities and resources both within and outside NASA. Aeronautics and Space Engineering Board (ASEB) member Charles Bolden, Jr. (an independent consultant), served as moderator; the panelists were Charles Kennel (Scripps Institution of Oceanography), Tamara Jernigan (Lawrence Livermore National Laboratory), and Lori Garver (The Avascent Group).

One panelist took the approach of examining the characteristics of two non-space-related examples of large, long-enduring federal research and development programs—high-energy physics and magnetically confined fusion research—and asked what lessons they might offer for space exploration. Common factors for the sustainability of those programs appeared to be driving motivation (such as the prestige associated with the pursuit of frontier scientific questions); capacity to produce visible, credible, incremental progress; programmatic balance (i.e., a portfolio of both large and small projects); geographic diversity; and international reach and collaboration. Participants acknowledged that the analogies are not perfect, especially with respect to the budgetary scale of space exploration compared with these examples in physics research.

The discussions in both workshop sessions highlighted factors that many participants saw as important to ensuring sustainability—realism about resources, leadership, relevance and value, and balance. These factors are summarized below.

Realism About Resources

Participants often returned to prior discussions about the perceived mismatch among expectations, resources, and program sustainability. One speaker noted that when responsibilities are expanded, as was the case when NASA embarked on the objectives of the Vision, additional resources are also needed or there can be impacts on other aspects of the agency's program. The speaker argued that trying to do everything with a flat budget was a "failed strategy."

Speakers cited both internal and external factors that can affect budget realism. Among the former are the impacts of project delays, the adequacy of contingency funds to cover technical risks, pressures for full employment at all NASA centers, and "management turbulence" that elicits defensive behavior throughout an organization when resources are seen as being both tight and threatened. External factors cited that might impact budget realism, and thereby program sustainability, included the capacity to respond to emerging policy drivers (e.g., rising competition from China and India and the emergence of concerns about climate and energy as major global issues), likely continued federal budget deficits, and potential congressional opposition to U.S. reliance on Russia during an extended launch hiatus after the space shuttle is retired.

Leadership

The concept of leadership played into discussions of sustainability in at least three ways. Some participants argued that strong leadership at the senior levels of NASA and the government is an essential factor in planning, articulating, and promoting NASA's program. In this context, some speakers viewed as a matter of considerable urgency the need for senior leaders to face up to what was described as a program that cannot be executed within the allotted budget. Second, there were calls for members of the space community, both inside and outside the government, to lead by establishing credible program cost estimates and carrying out programs within realistic budgets. Third, several participants argued that a space program that puts the United States in an international leadership role would have the greatest national impact and public support.

While some speakers expressed frustration over what was perceived to be an unrealistic current plan, some also posited that arguments can be made for increasing NASA's budget. One speaker recalled that in the mid-1980s and early 1990s, when NASA was trying to cope with the impacts of the Challenger accident and near-simultaneous failures of all major expendable launch vehicles, the agency was successful in securing additional funds that were important across multiple parts of NASA's program. Securing larger budgets required convincing arguments. This outcome was illustrated by a colloquy between two participants in which one suggested that the current situation is as if "we're a group of people having dinner and there wasn't enough food. Then we brought in a 300-lb visitor . . . . The best solution would be more food," to which a colleague replied, "And if your visitor were important enough, you might get it."

Participants offered diverse opinions about the relative leadership roles of NASA and the scientific community in achieving and operating within affordable budgets. Some noted that it has been common practice has been for NASA to ask the community to recommend the best science and then leave the determination of how to accomplish that science to NASA. Others cited experiences in which active participation by the outside community had led to important successes in bringing or keeping projects within realistic bounds. Examples of the latter included Earth science community efforts in the 1990s to define an affordable Earth Observing System program; Mars Science Laboratory science team efforts to bring the mission cost back within budget limits; and the recent National Research Council (NRC) assessment of candidates for Beyond Einstein program missions, a study that included engineers, managers, and cost experts working alongside scientists on the committee.[1]

International leadership as a central motivating factor for the civil space program was also a recurring theme. One speaker noted that a country might strive to exercise leadership in space for two different reasons—for prestige or for "techno-nationalism" (i.e., using technology to support national and geopolitical interests). Some speakers suggested that space-related efforts in many parts of the world have been influenced by what the United States undertakes; if the United States does it, others want to do so as well. Although there were differing opinions voiced about whether China poses a threat to U.S. leadership in space, several speakers argued that if the public saw China as having somehow taken the lead, there

---

[1] National Research Council, *NASA's Beyond Einstein Program: An Architecture for Implementation*, The National Academies Press, Washington, D.C., 2007.

would be an immediate reaction of alarm in the United States. As noted in Chapter 2, one speaker introduced the idea of the United States as a hegemon in space, that is, a nation that has been capable of exerting a dominant influence over others. Speakers suggested that the United States has an opportunity to choose to be the last unilateralist hegemon before the world goes global or to be a benevolent hegemon, leading the world in a collaborative effort on behalf of the international community.

Relevance and Value

Discussions about a rationale for the civil space program were characterized by frequent affirmations that space program proponents need to be able to articulate a clear and compelling purpose that communicates the value of investment in space activities to decision makers and the public. One speaker offered that, from a political perspective, the first thing that candidates for major national office ask is, "Why should I care?" There was considerable diversity of opinion about what kinds of answers to that question are most likely to motivate support for a sustainable program.

One panelist noted that, from a historical perspective, there have been five reasons for going into space: science, national security, commercial activities, a sense of human destiny and exploration, and national prestige and geopolitics. The first three of those reasons can be pursued without having to send people into space, whereas the last two have been factors in advocating on behalf of human spaceflight. The panelist concluded that, especially in view of the degree of public apathy about funding space programs, geopolitical arguments have been most important in sustaining human space exploration in the United States; no leader has wanted to go down in history as having been the one to cede this effort to others outside the United States.

Other speakers built on the historical perspective and argued that the geopolitical contributions of the space program can and should remain as central elements of the rationale for the future program. Participants described this kind of goal in two ways. Either the space program could be motivated by competition in which the United States commits to space exploration in order to respond to a perceived threat that others (e.g., China) might overtake the United States in space, or the United States could pursue a more cooperative approach in which the nation exerts geopolitical leadership to bring the international space community together to serve important global goals. The latter approach, which some speakers believed to have strong public staying power, would be the benevolent hegemon approach mentioned above.

Another panelist focused on domestic relevance, that is, on the need for the NASA program to demonstrate significant practical benefits and value to the public. When one asks the public about value that comes from investments in the National Institutes of Health, or the Federal Bureau of Investigation, or the Federal Aviation Administration, there are usually prompt answers. But there are no such ready answers with respect to NASA. The speaker argued that proponents of the Vision for Space Exploration[2] have failed to rise to the challenge of articulating the public value of space exploration because attention has been focused on the "how" (e.g., new launch systems hardware) instead of the "why." The speaker suggested that the case for space needs to be made on the basis of a combination of how space is emblematic of technological and scientific leadership and of how the investments in space projects have technological and economic benefits on Earth. Some participants agreed that it is important to be able to emphasize the relevance of space activities to such down-to-Earth concerns as climate change solutions, energy, breakthroughs in transportation, and diplomacy. Others cautioned that there are now many alternative avenues for technology investment (e.g., biotechnology and information systems) and that as a consequence, arguments for the power of space program spin-offs are no longer as compelling as they might have been in the past.

Another school of thought regarding what most interests the public about the value of space exploration was what some participants called the "wow factor." That is, people throughout the world are

---

[2] National Aeronautics and Space Administration, *The Vision for Space Exploration*, NP-2004-01-334-HQ, NASA, Washington, D.C., 2004.

still genuinely interested in and excited about what is accomplished in the space program. The Hubble Space Telescope and the Mars Exploration Rovers were cited as examples. Some noted that the public does connect in a different way when people—that is, astronauts—are physically involved in the activities. Others added that if the fledgling commercial space tourism industry gains traction there will be enormous new interest in space. During these discussions some participants raised two cautionary notes. First, there were participants who argued that returning to the Moon, which constitutes the first major near-term objective of the Vision, is simply not exciting to people or viewed as expansive or creative. Second, several participants indicated that today's young generation has different expectations that include a strong desire for a participative experience. According to these speakers, the degree of participation and virtual presence that today's youth have come to expect is lacking in the Vision.

Balance

Several participants made comments about how the perceptions of balance can impact sustainability, and they noted that the concept of balance has multiple dimensions that not everyone will see in the same way. One speaker suggested that at the highest level, balance should be judged by how well program elements are deployed to respond to national needs, such as addressing national security, science, global health, education, and workforce development, all of which might be characterized as high-risk, high-reward areas. Another speaker noted that policy makers assess balance in terms of how program priorities are prescribed by the law (National Aeronautics and Space Act of 1958, Public Law 85-568[3]) and how they support priorities set by national policy, noting that electing policy makers and having them make decisions about top-level priorities has worked in the space program for 50 years. The speaker argued that the current administration's space policy[4] has seven top-level goals, which are about equally divided between items for civil space programs and for national security and defense space programs. The former can all be linked to NASA's strategic plan.

One speaker noted that there is an unfortunate tone in discussions of balance because they appear to pit human space exploration against the scientific and aeronautics sides of NASA. This kind of debate, the speaker indicated, stems from the fact that all sides think that resources for their programs are scarce. The balance deliberation should really be about how to achieve synergies among different parts of NASA, among NASA and other agencies and countries, and between NASA and the private sector.

That speaker also suggested that good metrics for proper balance would include clear indications of measurable progress and success as well as attention to the long-term health of the underlying disciplines. The speaker noted a need to balance long- and short-term goals to prepare the country for unforeseen future developments as well as demonstrating progress toward nearer-term goals. Another speaker suggested that assessing balance requires recognizing that different elements of NASA's program have different constituencies with different objectives or values. For example, the scientific community measures success in terms of the degree of scientific progress in pursuing its decadal survey priorities;[5] the Earth science community measures progress against how well it can address national environmental needs; and the aeronautics community measures progress in terms of responding to the needs of commercial and military air transport interests. Human spaceflight is in a different situation in that its constituency should be the public, although its focus seems to be on NASA rather than the public as a constituency now.

---

[3] National Aeronautics and Space Act of 1958, Public Law 85-568, 72 Stat., 426, July 29, 1958, Record Group 255, National Archives and Records Administration, Washington, D.C; available in NASA Historical Reference Collection, History Office, NASA Headquarters, Washington, D.C.

[4] Office of Science and Technology Policy, *U.S. National Space Policy*, National Security Presidential Directive 49, unclassified version released on October 6, 2006, available at http://www.ostp.gov/galleries/default-file/Unclassified%20National%20Space%20Policy%20--%20FINAL.pdf.

[5] For more information about decadal surveys, see National Research Council, *Decadal Science Strategy Surveys: Report of a Workshop*, The National Academies Press, Washington, D.C., 2007.

Another speaker summed up the discussions of balance by noting that when one goal dwarfs all others and keeps them from being achieved, there is a balance problem. Adjusting such an imbalance takes time. The speaker noted that one can never do enough for balance, but hard choices need to be made and decision makers just have to make the best of it.

EARTH OBSERVATIONS

SSB member Jack Fellows (University Corporation for Atmospheric Research) moderated the session on civil government missions in Earth observations, and panelists Johannes Loschnigg (NRC staff consultant and former member of the staff of the House Subcommittee on Space and Aeronautics), Berrien Moore III (University of New Hampshire), and Soroosh Sorooshian (University of California, Irvine) opened the discussions. Like the sessions that preceded it, the overall tone of this session was pessimistic.

Speakers summarized evidence for significant climate-related changes that are being measured now—for example, the accelerating rate of carbon emissions from the burning of fossil fuels, estimates that nearly half of all current carbon dioxide (CO_2) emissions are not removed by natural sinks but stay in the atmosphere, and the acceleration of atmospheric CO_2 concentration since the late 1960s—and argued that very serious global climate change problems are likely to continue for the next 50 years or so. The issues were described as being research problems rather than operational problems, but speakers noted that federal funding for the National Oceanic and Atmospheric Administration (NOAA, the operational agency) has been increasing while NASA (the research agency) has experienced decreasing funding. One speaker described this situation as "a perfect storm."

A speaker in an earlier session had argued that a critical need is for the United States to rebuild its leadership capacity in Earth science. Recognizing that NASA Earth observations are embedded in both national and international programs, the speaker posed several strategic questions, as follows:

- How should NASA participate in the national Earth science program?
- How can NASA regain interagency and international high ground?
- How can NASA regain the technical high ground?

Responding, in part, to these questions and the questions posed for the session panelists (see Appendix B), one speaker suggested that the basic division of labor should continue to be that NASA does the scientific and technological research and then hands the results off to NOAA for applying them to operational responsibilities. Other speakers agreed that the two agencies should retain their currently assigned roles. However, the two agencies have distinctly different cultures that influence their priorities. The Tropical Rainfall Measurement Mission was cited as an example of a transition failure. Because its research phase had ended, NASA wished to terminate the mission or hand the mission over to NOAA, but NOAA did not wish to assume the operational costs in spite of the mission's operational value.[6] Similarly, the restructuring of the National Polar-orbiting Operational Environmental Satellite System broke down in the sense that the Department of Defense drove the restructuring, while NOAA retreated to those elements that supported its core weather-monitoring and weather-forecasting mission and NASA ended up not being much of a player in decisions in spite of its research needs. One participant noted that when considering interagency activities, it is important to recognize that NASA and NOAA have distinctly different cultures. Another participant noted that as a general rule it is not a practical idea to fly research instruments on operational missions.

---

[6] National Research Council, *Assessment of the Benefits of Extending the Tropical Rainfall Measurement Mission: A Perspective from the Research and Operations Communities, Interim Report*, The National Academies Press, Washington, D.C., 2006.

One speaker noted that the 2006 national space policy included four items for action by NASA and NOAA,[7] but that implementation of those actions had not succeeded. The speaker observed that interagency activities work well in an environment of increasing budgets, but not when resources are falling. Clear authority and direction are necessary in the latter case.

CAPABILITIES AND INFRASTRUCTURE

The final session on strategic issues, led by ASEB chair Raymond Colladay (Lockheed Martin Astronautics, retired), addressed questions regarding future needs and gaps in capability and infrastructure. The panelists were John Klineberg (Space Systems/Loral, retired), Thomas Zurbuchen (University of Michigan), and Ian Pryke (George Mason University).

In contrast to the concern that marked much of the earlier discussion during the workshop, some participants in this session suggested that there is reason for optimism. That is, if there are well-articulated national priorities and the resources to back them up, then market forces will ensure that industry will be able to provide the necessary capabilities and infrastructure.

A concern receiving considerable attention related to the challenge of interesting and recruiting young people who will constitute the high-technology workforce that will carry out the space exploration program. One speaker described experience with very bright university students who initially selected but then opted out of aerospace careers because there were too few opportunities for substantive hands-on work compared with more exciting and challenging alternatives in other career fields. The speaker cited three factors that contribute to this problem: the lack of a sense that space programs offer an opportunity to make an important impact, the lack of effective partnerships between universities and NASA that focus on creating talent, and the fact that the International Traffic in Arms Regulations keeps some of the best non-U.S. students out of the pipeline.

Another speaker addressed the recruitment issue by citing the results of market surveys that examined public interest in the space program.[8] The number of 18- to 24-year-olds who indicated that they were excited by or interested in returning humans to the Moon (35 percent) was far smaller than the number of 45- to 64-year-olds (80 percent). Furthermore, in the 18- to 24-year-old age group there was a 3-to-1 margin of opposition to sending humans to Mars, but slightly more than half of the people sampled in that age group indicated support for more robotic missions to Mars. The latter group showed special interest in opportunities for telepresence, that is, missions that could support a remotely interactive or virtual presence on Mars. The speaker concluded that it is essential that this age group be cultivated so that it will not only be supportive of space exploration but will also create the pool of future program leaders.

In subsequent discussions some participants repeated support for the ideas that young people can become interested in space exploration in general when they see opportunities to become engaged through contemporary media that create a virtual presence, and that young people can become interested in space exploration careers when they see opportunities for real hands-on work in the program. Others noted that there is evidence of contemporary success along these lines, and they cited two institutions where turnover rates are very low—namely, at the Jet Propulsion Laboratory and at Lockheed Martin Corporation.

---

[7] Office of Science and Technology Policy, *U.S. National Space Policy*, National Security Presidential Directive 49, unclassified version released on October 6, 2006, available at http://www.ostp.gov/galleries/default-file/Unclassified%20National%20Space%20Policy%20--%20FINAL.pdf.

[8] See P. Finarelli and I. Pryke, *Building and Maintaining the Constituency for Long-term Space Exploration*, Center for Aerospace Policy, George Washington University, Washington, D.C., available at http://www.gmupolicy.net/aerospace/IAC-PaperFINAL.pdf.

4

Epilogue

At the final session of the 2007 workshop, moderator Raymond Colladay (Lockheed Martin Astronautics, retired), Aeronautics and Space Engineering Board chair, invited all participants to provide concluding observations. These are summarized below in terms of three broad categories.

COMMUNICATING ABOUT SPACE EXPLORATION

About a quarter of the workshop participants cited aspects of how to communicate with the public and policy makers about space exploration and about what that message should be. These perspectives began with the view that the various elements of NASA's program—human spaceflight, space and Earth science, and aeronautics—should not be cast as separate, competing elements but should be seen as integrated in a perspective that should lead to a single core mission for NASA—"to inspire." Others added that there needs to be a new paradigm for motivating human space exploration, because the Vision for Space Exploration[1] as it is now presented fails to engage the public. A key to this argument was that to date there has been too much emphasis on hardware and on "how" the Vision will be pursued rather than on the more important question for the public of "why" we will explore. Speakers also noted that being able to communicate examples of steady progress along the way is very important. Finally, some participants added that emphasizing Chinese lunar mission plans and progress will not be effective in motivating the public because many people will say that the United States has already accomplished that goal.

Speakers who addressed the question of why there should be a space program focused on the idea that space exploration is emblematic of leadership. Space, according to these viewpoints, provides a powerful vehicle for innovation and for pursuit of technological and economic benefits. Participants also argued that NASA is primed to make major contributions on behalf of the United States in addressing pressing issues about global climate change.

Most comments about the communications challenge emphasized that the message has to connect with today's younger generation. To succeed at that, participants argued, one must listen to the interests of young people and to young professionals. For example, speakers cited the idea of providing streaming images from Mars and giving people a virtual presence experience on Mars or elsewhere in space. Some participants also reminded their colleagues that "space is cool" and that this notion can still generate public interest.

INTERNATIONAL COMPETITION, COOPERATION, AND LEADERSHIP

Another quarter of the 2007 workshop's participants commented that the importance of international leadership and cooperation stood out as being particularly noteworthy. Several argued not

---

[1] National Aeronautics and Space Administration, *The Vision for Space Exploration*, NP-2004-01-334-HQ, NASA, Washington, D.C., 2004.

only that international aspects of space exploration are very important but also that a geopolitical context for the U.S. space program is essential. One speaker posited that the prevailing view needs to be that "space is not a race but a responsibility"—that is, that the principal aim of the U.S. space program should be international leadership in exploration, science, and technology and that the United States should use space activities to lead efforts on global issues such as energy, resources, and climate, that is, to become a benevolent hegemon.

Some speakers cited NASA's science programs and the International Space Station (ISS) as notable international cooperation success stories. One speaker opined that successful human missions to Mars can only be accomplished through international cooperation. Some participants cited problems arising from the current application of export controls, especially the International Traffic in Arms Regulations (ITAR), which create impediments to much international cooperation on space projects, and argued that these impediments need to be fixed.[2]

ENSURING ROBUSTNESS THROUGH NEW APPROACHES AND ATTITUDES

Nearly half of the workshop participants mentioned impressions about threats to the robustness of NASA's program and commented about potential approaches to cope with those problems. Several speakers urged that the first step is for senior government leaders to acknowledge that there is a problem and that, as is noted in Chapter 2, there is a serious mismatch between NASA's assigned responsibilities and its available resources—that is, issues of budget realism, program feasibility, and sustainability must be addressed head-on. Others added that this problem needs to be specified better quantitatively.

Speakers offered a number of ideas about what is needed to deal with the current problems. Elements of a strategy included putting a budget wall between resources for science and exploration, becoming more open to international and commercial partnerships, protecting investments in capabilities that will be needed in the future, and limiting lunar mission activities to only what is needed to prepare for future Mars missions. Some participants argued for planning an exploration program that goes directly to Mars and bypasses the Moon.

One speaker advocated a specific strategy for avoiding the impending "train wreck" to which many other participants referred during the workshop. The speaker argued that there is a need to "slow down the train" by deferring the first human mission to the Moon; extending the use of the ISS in support of R&D for later human exploration; establishing telepresence on the Moon; creating an environment of institutional stability in NASA's program elements; building globally inclusive working groups on direct missions to Mars, global change, and space science; and defining real, meaningful jobs for humans in space.

Finally, a few participants commented that the workshop discussions had been too pessimistic and that there is reason for optimism, especially for the long term. These participants argued that there is great promise for long-term progress in much of NASA's program and that if there is a willingness to make some changes in course, the program will succeed.

In closing the final session, moderator Colladay remarked that at the beginning of the workshop he had urged people to take a non-advocate approach and to look at the space program in a broad context. He observed that, in fact, people had looked at the current situation more critically than is often the case in other industries. The discussions focused more on problems than on solutions, but he suggested that the first step toward solving problems is to engage with the problem. He concluded that the workshop succeeded in doing that.

---

[2] For a more thorough discussion of the impacts of ITAR on space science, see National Research Council, *Space Science and the International Traffic in Arms Regulations: A Workshop Summary*, The National Academies Press, Washington, D.C., 2008.

Appendixes

A

Statement of Task

An ad hoc committee under the auspices of the Space Studies Board, working in collaboration with the Aeronautics and Space Engineering Board, will organize a public workshop for the purpose of encouraging broad national discussion about future directions of the U.S. civil space program. The workshop will utilize invited talks, panel discussions, and general discussions to review developments since the two boards held a similar workshop in 2003 and will revisit aspects of the question "What are the principal purposes, goals, and priorities of the U.S. civil space program?"

Among the ancillary questions that could be open for discussion are the following:

1. What are the fundamental purposes of the U.S. space program and what are the roles and relationships for space activities to promote national security, societal benefits, scientific and technological advancement, commercial and economic benefits, and international relations?
2. What are the appropriate roles of the federal government vis-à-vis the private sector?
3. How can expansive expectations for the total content of the civil space program be reconciled with realistic expectations for total program resources?
4. What is required to ensure that national goals for human space exploration are sustainable?
5. What are the relationships between U.S. national space goals and those of other countries, and where are there current and future opportunities for cooperation and synergism?

The goal of the workshop will not be to develop definitive answers to any of these questions but to air a range of views and perspectives that will serve to inform later broader public discussion of such questions and a prospective comprehensive study on U.S. space policy.

PRELIMINARY WORK PLAN

The organizing committee will plan and hold a one-and-one-half-day public workshop in tandem with a scheduled meeting of the Space Studies Board on November 29-30, 2007, at the Beckman Center. Approximately 12 outside participants will be invited to make presentations and join in panel discussions during the workshop. Overall participation in the workshop will include members of the SSB and ASEB, other experts from academia and industry, and representatives from relevant federal agencies and Congress. A workshop summary (type 3) report will be prepared by an appointed rapporteur with the assistance of staff. The report will summarize what occurred at the workshop and will include summaries of individual presentations, but it will not present consensus conclusions or recommendations. The report will be published within four months of the workshop. It will serve to inform later broader public discussion of such questions and a prospective comprehensive study on U.S. space policy.

B

Workshop Agenda

NOVEMBER 29, 2007

1:30 p.m.　　Welcome, Introductions, Workshop Objectives
　　　　　　　R. Colladay and L. Fisk

2:00　　　　　Situational Assessment
　　　　　　　Moderator: T. Young; Panelists: B. Alexander, F. Harrison, J. Zimmerman

　　　　　　　Key Changes and Developments Since 2003, such as the following:

- Confronting a fundamental lack of financial robustness in the overall civil space program,
- Progress to date and challenges ahead for the Vision for Space Exploration,
- Emergence of China as a space contender as other international players also continue to become more independent and competitive,
- NPOESS and GOES-R program crises in U.S. Earth observations program,
- Evolution in public and political views about climate change, and
- Budgetary and political developments and their impact on the current environment.

3:15　　　　　Break

3:30　　　　　National and International Context for Space
　　　　　　　Moderator: C. Bennett; Panelists: G. Gugliotta, J. Johnson-Freese, R. Launius

- Are the expectations of space program advocates out of step with reality with regard to NASA's position in the national agenda?
- Where does NASA sit in the larger national and international context?
- How important are civil space activities to broad national goals to promote national security, societal and cultural benefits, scientific and technological advancement, commercial competitiveness and economic benefits, and international relations?
- What are the relationships between U.S. national space goals and those of other countries, and where are there current and future opportunities for cooperation and synergism?
- How important are the stated intentions of China and Russia for exploitation of the Moon to U.S. space exploration?

6:00　　　　　Reception and Dinner

NOTE: See Appendix C, "Workshop Participants," for the full name and affiliation of each moderator and panelist.

NOVEMBER 30, 2007

8:30 a.m.	Sustainability Issues and Options for Solutions: Affordability, Public Interest, and Political Will *Moderator: J. Pawelczyk; Panelists: P. Carliner, G. Paulikas, R. Truly, G. Whitesides*

- How can expansive expectations for the total content of the civil space program be reconciled with realistic expectations for total program resources?
- What is required to ensure that national goals for human space exploration are sustainable?
- Are there proven strategies for ensuring sustainability for large federal programs?

10:15	Break
10:30	Balance Issues and Options for Solutions *Moderator: C. Bolden; Panelists: T. Jernigan, C. Kennel, L. Garver*

- How should decision makers assess an appropriate balance between NASA's programs in human spaceflight vs. science vs. aeronautics?
- Is "balance" the same as "investment portfolio mix"?
- What are appropriate criteria or metrics for achieving "balance"?
- Roles of NASA vs. roles of others
  —What are the appropriate roles of NASA vis-à-vis other government agencies?
  —What are the appropriate roles of the federal government vis-à-vis the private sector?

12:15 p.m.	Lunch
1:30	Civil Government Missions in Earth Observations *Moderator: J. Fellows; Panelists: J. Loschnigg, B. Moore, S. Sorooshian*

- What should be NASA's role in helping the National Oceanic and Atmospheric Administration (NOAA) acquire the data needed to assess global climate change?
- What are the appropriate roles and responsibilities of NASA, NOAA, and other agencies in Earth observations from space?

2:15	Capabilities and Infrastructure *Moderator: R. Colladay; Panelists: J. Klineberg, T. Zurbuchen, I. Pryke*

- Are there critical unmet needs or anticipated gaps that should be addressed to give the U.S. the capability to achieve its civil space goals, and what strategies are needed to meet expected long-term needs?
  —U.S. space industrial base, NASA centers, and academia
  —Access to space
  —Technology development

3:15	Break
3:30	Synthesis and Wrap-up: Summary Comments by All Participants *Moderator: R. Colladay*

C

Workshop Participants

SPACE STUDIES BOARD MEMBERS

Lennard A. Fisk, *Chair,* Thomas M. Donahue Distinguished Professor, Department of Atmospheric, Oceanic and Space Sciences, University of Michigan
A. Thomas Young, *Vice Chair,* Executive Vice President (retired), Lockheed Martin Corporation
Spiro K. Antiochos, Head, Solar Theory Section, Space Science Division, Naval Research Laboratory
Daniel N. Baker, Professor and Director, Laboratory for Atmospheric and Space Physics, University of Colorado
Steven J. Battel, President, Battel Engineering
Charles L. Bennett, Professor of Physics and Astronomy, Johns Hopkins University
Elizabeth R. Cantwell, Deputy Division Leader, Space and Response Division, Los Alamos National Laboratory
Alan Dressler, Astronomer, Observatories of the Carnegie Institution
Jack D. Fellows, Vice President of Corporate Affairs and Director of the UCAR Office of Programs, University Corporation for Atmospheric Research (UCAR)
Fiona A. Harrison, Professor of Physics and Astronomy, California Institute of Technology
Tamara E. Jernigan, Principal Deputy Associate Director, Physics and Advanced Technology, Lawrence Livermore National Laboratory
Klaus Keil, Professor, Hawaii Institute of Geophysics and Planetology, School of Ocean and Earth Science and Technology
Molly K. Macauley, Senior Fellow, Resources for the Future, Inc.
Berrien Moore III, Director, Institute for the Study of Earth, Oceans and Space, University of New Hampshire
James A. Pawelczyk, Associate Professor of Physiology, Kinesiology and Medicine, Pennsylvania State University
Soroosh Sorooshian, Distinguished Professor of Civil and Environmental Engineering, University of California, Irvine
Richard H. Truly, Director (retired), National Renewable Energy Laboratory
Joan Vernikos, Consultant, Thirdage, LLC
Charles E. Woodward, Associate Professor of Astronomy, University of Minnesota
Gary P. Zank, Director, Institute of Geophysics and Planetary Physics, University of California, Riverside

AERONAUTICS AND SPACE ENGINEERING BOARD MEMBERS

Raymond S. Colladay, *Chair,* President (retired), Lockheed Martin Astronautics
Charles F. Bolden, Jr., Chief Executive Officer, Jack and Panther, LLC
John M. Klineberg, President (retired), Space Systems/Loral

INVITED PARTICIPANTS

Bretton Alexander, Executive Director, X Prize Foundation
Gale Allen, Deputy Director of Strategic Integration and Management, NASA Headquarters
Marc S. Allen, Assistant Associate Administrator for Strategy, Policy and International, Science Mission Directorate, NASA Headquarters
Steven Beckwith, Director Emeritus, Space Telescope Science Institute
Steven Benner, Distinguished Fellow, Foundation for Applied Molecular Evolution
Jacques Blamont, Professor of Physics, University of Paris
Roger Bonnet, Director, International Space Sciences Institute
Blake Bullock, Mission Integration Manager, Northrop Grumman Space Technology
Paul Carliner, Independent Consultant
John Casani, Senior Advisor to the Director, Jet Propulsion Laboratory
Alphonso V. Diaz, Vice Chancellor for Administration, University of California at Riverside
Edward G. Feddeman, Senior Staff Member, House Subcommittee on Space and Aeronautics, Committee on Science and Technology
Lori B. Garver, Senior Advisor, The Avascent Group
Daniel S. Goldin, Chairman and Chief Executive Officer, The Intellisis Companies
Guy B. Gugliotta, Journalist and Author
Gerhard Haerendel, Professor, Max Planck Institute, Garching, Germany
Joan Johnson-Freese, Professor of National Security Decision Making, Naval War College
Charles F. Kennel, Professor, Scripps Institution of Oceanography, University of California, San Diego
Roger D. Launius, Chief Historian, National Air and Space Museum
Matt Mountain, Director, Space Telescope Science Institute
Richard Obermann, Staff Director, Subcommittee on Space and Aeronautics, House Committee on Science and Technology
George A. Paulikas, Executive Vice President (retired), The Aerospace Corporation
Angela Phillips-Diaz, Director, Strategic Communications and Development Directorate, NASA Ames Research Center
Ian W. Pryke, Senior Fellow/Assistant Professor, Center for Aerospace Policy Research, School of Public Policy of George Mason University
Amy Scott, Senior Federal Relations Officer, Association of American Universities
George T. Whitesides, Executive Officer, National Space Society
Jennifer Wiseman, Chief of the ExoPlanets and Stellar Astrophysics Laboratory, NASA Goddard Space Flight Center
James V. Zimmerman, President, International Space Services, Inc.
Thomas H. Zurbuchen, Associate Professor of Space Science and Engineering, University of Michigan

NATIONAL RESEARCH COUNCIL STAFF

Barbara S. Akinwole, Information Management Associate, Space Studies Board (SSB)
Joseph K. Alexander, Senior Program Officer, SSB
Carmela J. Chamberlain, Program Associate, SSB
Arthur A. Charo, Senior Program Officer, SSB
Dwayne A. Day, Senior Program Associate, SSB
Brian D. Dewhurst, Senior Program Associate, Board on Physics and Astronomy (BPA)
Sandra J. Graham, Senior Program Officer, SSB
Johannes Loschnigg, Consultant, SSB
Celeste Naylor, Senior Program Assistant, SSB
Tanja E. Pilzak, Administrative Coordinator, SSB

Robert L. Riemer, Senior Program Officer, BPA
Christina O. Shipman, Financial Associate, SSB
David H. Smith, Senior Program Officer, SSB
Kerrie Smith, Program Officer, Aeronautics and Space Engineering Board (ASEB)
Marcia S. Smith, Director, SSB and ASEB
Victoria Swisher, Research Associate, SSB